Saudi Arabian REVENUES AND EXPENDITURES

The Potential for Foreign Exchange Savings

DONALD A. WELLS

Published by Resources for the Future, Inc.
Distributed by The Johns Hopkins University Press
Baltimore and London

Saudi Arabian Revenues and Expenditures

Resources for the Future is a nonprofit corporation for research and education in the development, conservation, and use of natural resources and the improvement of the quality of the environment. It was established in 1952 with the cooperation of the Ford Foundation. Part of the work of Resources for the Future is carried out by its resident staff; part is supported by grants to universities and other nonprofit organizations. Unless otherwise stated, interpretations and conclusions in RFF publications are those of the authors; the organization takes responsibility for the selection of significant subjects for study, the competence of the researchers, and their freedom of inquiry. The manuscript was edited by Evelyn Pettit and Joan Tron.

RFF editors: Mark Reinsberg, Joan R. Tron, Ruth B. Haas, Margaret Ingram

Manufactured in the United States of America

Library of Congress Catalog Card Number 74-19945

ISBN 0-8018-1689-0 Paper $3.00

Table of Contents

PREFACE v

ACKNOWLEDGMENTS ix

1. INTRODUCTION 1

2. ESTIMATES OF FUTURE PETROLEUM REVENUES 4
   - Production Levels 4
   - Prices 5
   - Revenues 6

3. GOVERNMENT REVENUES AND EXPENDITURES: THE 1960s EXPERIENCE 7
   - Revenues 7
   - Appropriations and Expenditures 7
     - Appropriations 9
     - Consumption Expenditures 14
     - Development Expenditures 15

4. REVENUES AND EXPENDITURES, 1973–1980 18
   - Revenues 18
   - Expenditures 20
     - Factors Affecting Expenditures 23

5. SAVINGS 28
The Implications of Saudi Foreign Exchange Savings 28
The Composition of External Assets 29
Equity Investments 30
Short-term Assets 32
The International Exchange Rate System 33

NOTES 34

List of Tables

1. Revenues, Annual Rates of Change: 1959–72. 8
2. Revenues and Appropriations 1959–73, with Patterns of Expenditure 1964–72. 10
3. Revenues, Appropriations, and Expenditures: Annual Rates of Change, 1960–72. 12
4. Revenues, Projected Annual Rates of Change, 1974–80. 19
5. Revenues, Expenditures, and Savings, Projected, 1973–80. 29
6. External Assets, 1960–73. 30

Preface

Perhaps it is just as well that the research community does not concern itself with thinking up all the potential questions to which answers may some day be needed. Some rule of relevance and timeliness does seem to order the agenda. But at times events move so rapidly that scholars are left far behind, unable to come up with good answers to burning questions. Conjecture, speculation, and the imaginative drawing of analogies then take the place of careful research.

The Arab oil embargo and the drastic increases in price associated with it are such events. They have caught the scholarly community with its cupboard bare, or at best only spottily occupied. Among the many questions to which answers are outstanding is the degree to which the impact of the rapid and large accumulation of oil revenues will affect monetary and eventually economic stability the world over. While new jargon has quickly emerged and one talks glibly about "petrodollars" and of "recycling" the swollen treasuries of the oil-exporting countries, the ice on which the discussants tread is pretty thin.

An important aspect of the debate is the capacity of the Arab nations to absorb the revenues through expanding consumption by their citizens as well as their investment in productive facilities.

"Absorptive capacity" is a concept long employed among students of foreign aid and development economics generally. It relates to the degree to which a country can, in an orderly manner, utilize an inflow of funds. In the case of some of the Middle Eastern oil-producing countries the rise in the volume of such funds expected to flow in their direction is so steep that neither the existing literature nor recourse to the country's own experience is of much help for establishing a firm base on which to build forecasts. Yet, the volume of the "loose cash" that will emerge as the residual after domestic absorption and that will be looking for employment outside the exporter countries will be a critical factor in judging prospects for world economic stability.

Donald Wells wisely anticipated the importance of the issue many months before the war of October 1973. Indeed, he had completed a first draft before the war broke out. This research report represents a revised version that attempts to take account of subsequent developments in the area of prospective price and production development. Little had to be changed, however, that relates to absorptive capacity. Unhappily, the reason is that the data on which to base judgement have remained as sparse as they were before, giving few indications as to how funds will be employed.

Wells has rightly chosen Saudi Arabia as the key country, given the contrast between the vast potential for expanding oil production and the modest economic stage of development. Saudi Arabia, that is, may well be the key exhibit. In developing his thesis that Saudi Arabia's absorptive capacity is greater than the casual observer would suspect, Wells leans heavily on an analysis of past budget appropriations and expenditure records and on evidence of unmet needs of the country's population. I use the term "needs" advisedly, since in the absence so far of the counterpart income these needs have not attained the dimension of "demand." This, however, is the transformation that Wells expects in the future, largely through governmental intermediation via enlarged consumer goods imports, social services, housing, and—last but not least—military hardware. In the investment area, agricultural infrastructure and oil-related industry rank high as potential outlets.

The net result is a volume of funds seeking employment abroad that is large but not overwhelming. To the extent that this is so because a large military buildup sops up funds, it is hard to know whether or not to characterize the outlook as salutary. But that is not Wells's task. What he has done is put on the table a substantial volume of data and supplied ways of drawing conclusions from them. It will be up to the reader to judge their relevance as well as their usefulness for further analysis. We owe Wells a debt of gratitude for having opened one door a little wider in an area where most doors are still closed, and RFF is glad to have funded the research.

Hans H. Landsberg
Director, Energy and
Minerals Program

Washington, D. C.
July 1974

Acknowledgments

I would like to thank: Resources for the Future for its support of this research; Sam Schurr for his early encouragement; Faisal Al-Bashir for research assistance; M. A. Adelman, James Akins, Helmut Frank, Charles Issawi, Hans Landsberg and Raymond Mikesell for their helpful comments and suggestions from a reading of an earlier draft; Joan Tron for editorial assistance.

1

Introduction

Because of its present stage of development and the rapid rise in petroleum income which it will experience over the 1973–80 period compared with previous income levels, Saudi Arabia will most likely have more discretion than other major exporters in disposing of future earnings. By estimating future revenues, analyzing past expenditure levels and patterns, and assessing potential development strategy, this study attempts to determine the capacity of Saudi Arabia to "absorb" income and to spend it in a manner consistent with national objectives. To the extent that earnings exceed this capacity, Saudi Arabia will save foreign exchange, and the greater the level of these savings the more Saudi Arabia will emerge as a major force in international financial markets.

Until the recent war in the Middle East and the subsequent disruption of output and deliveries, there was general agreement among the experts about ,the potential supply of and demand for petroleum over the 1973–80 period. This study analyzes the consequences of Saudi Arabia's resuming deliveries and responding to the market considerations which formed the basis for the earlier forecasts.

Because the volume of production and the price of petroleum over the 1973–80 period cannot be predicted with much precision, an attempt is made to determine a range over which production and prices might vary. Some judgments are offered as to which estimates appear to have the most validity.

This study proposes to estimate (1) a range for the volume and value of Saudi Arabian petroleum exports annually, 1973–80; (2) Saudi government petroleum revenues from royalty and tax payments, and revenues from the Saudi participation share of petroleum output; (3) the propensity to save out of foreign exchange receipts during the 1973–80 period, by analyzing government expenditures for consumption and development; and (4) the levels of foreign exchange savings for the same period.

This study does not attempt to assess the likelihood or the consequences of Saudi Arabia's restricting petroleum output to relatively low levels, e. g., 10 million barrels per day (b/d) by 1980. The Saudi authorities might restrict production because of international political considerations; out of concern that a too-rapid increase in wealth would have adverse effects on their society; or because they concluded that oil in the ground would appreciate in value more rapidly than returns to income from currently produced oil. Whatever the reasons, if Saudi revenues are restricted, the principal implication for this study is that the lower level of revenues will minimize the potential international financial impact of Saudi Arabian earnings; with a relatively lower rate of increase of earnings, Saudi Arabia is not likely to accumulate large foreign exchange balances. This study focuses on the implications of major increases in Saudi revenues for foreign exchange holdings. While production limitations would have serious and deleterious consequences for the energy requirements of industrialized countries, these considerations are outside the scope of this paper.

My principal sources of background data are the *Annual Reports* of the Saudi Arabia Monetary Agency (SAMA), the Central Department of Statistics of the Ministry of Finance and National Economy, and the Central Planning Organization. Faisal Saffoq al-Bashir, in "An Econometric Model of the Saudi Arabian Economy: 1960–1970" (unpublished Ph. D. dissertation, University of Arizona, 1973) has compiled data from various official sources, which represent the most complete set of statistics on those sectors of the economy he analyzes.

No consistent time series data exist for the 1960–72 period, and the sources frequently give conflicting data without explanation. While the data are not as reliable as I would prefer, I believe that they form an adequate base for the analysis here. The recent data are more reliable than those for the early 1960s, although some of the data for 1971 and 1972 are not entirely from official sources, and official data may be subsequently revised. Data involving the petroleum sector and the foreign sector are the most reliable. In the area of government expenditures, data on appropriations are better than those on actual expenditures. Budget data for the early 1960s, as published by the Saudi government, are neither complete nor firm. During the fiscal years for which they were prepared, the budgets were at times changed by royal decrees; these changes were often not incorporated officially into the published budget. As budgetary procedures have been implemented and become regularized, the data have become better indicators of appropriations and programs.

All data are presented by calendar year. Many Saudi sources use the Islamic calendar, which is based on the lunar year (354 days with eleven leap years of 355 days every cycle of thirty years); the first year of this calendar is the year that the Prophet Muhammad emigrated from Mecca to Medina (A. D. 622). The Saudi fiscal year of 1389/90, for example, corresponds to Gregorian calendar dates September 13, 1969, to August 31, 1970. Statistics are presented in the Gregorian calendar year by approximating the Islamic year; for example, 1389/90 is presented as 1970.

In converting riyal amounts to U.S. dollar amounts, I used the following official exchange rates:

1959	$1	= 3.750 riyals
1960–71		= 4.500
1972		= 4.145
1973		= 3.730

These correspond to the market rate for each period. The riyal equivalent of dollar revenues during the 1970s will of course vary with the market rate of exchange.

2

Estimates of Future Petroleum Revenues

The determination of the Saudi Arabian government's petroleum revenues results from estimates of production levels, participation shares, prices, and taxation rates.[1] There is no consensus as to the likely values of these variables by 1980. Estimates vary widely. For this reason, it is preferable to accept some range of estimates and to establish the levels of government revenues and foreign exchange savings associated with the individual assessments within the range.

Production Levels

By August 1973 Saudi Arabia's output averaged 8.5 million barrels per day (b/d).[2] With the announcement on October 18, 1973, that Saudi production would be cut by 10 percent, and with further reductions being imposed and then eased, it has become difficult to anticipate production levels in either the near or more distant future. A *Petroleum Intelligence Weekly* survey of world petroleum production and expansion programs through 1973 had indicated planned crude oil capacity increases centered on the program of Aramco in Saudi Arabia; by the end of 1975 Aramco planned to increase its capacity to 11.6 b/d, but whether or not Saudi Arabia will be producing that amount remains conjectural. Estimates of the period beyond, from 1976 to 1980, are highly speculative. However, assuming that production and trade will respond to market conditions, it is estimated that Saudi Arabian production of petroleum in 1980 might range from 14 to 22 million b/d. This range is based on estimates made in the press and other sources before October 1973. For example, a report of the U.S. Senate Subcommittee on the Near East estimated a level of 14 million barrels per day.[4] James E. Akins, writing in the April 1973 *Foreign Affairs,* used an estimate of 20 million barrels per day, based on the goal set by Minister of Petroleum Ahmad Zaki Yamani, although Akins thought this level ". . . seems improbably high."[5] A May 1973 article in *The Economist* used a range of 10–20 million b/d by 1980, but stated that "the most likely estimate is that production will level out at between 12 million and 15 million barrels a day."[6] Relying on the American Universities Field Staff, Inc., a report in *Oil and Gas Journal* projected 1980 production at 20 million b/d.[7] Since Aramco planned capacity increases of approximately 80 percent in the 1973–75 period, with total capacity reaching 11.6 million b/d by the end of 1975, an additional doubling of total capacity over the following five-year period would not be unreasonable. The only major restraint on a production level of 20 to 22 million b/d in 1980 seems to be a decision by the Saudis to limit production; although in the face of high prices, efforts at energy conservation by consuming nations could restrict Saudi Arabian production below this level, or the Saudis might choose to limit production for other reasons.

Prices

Based on announcements by the Organization of Petroleum Exporting Countries (OPEC) in October and December 1973, the market price of Arabian light crude was increased considerably. At the end of 1973 the tax reference, or posted, price was set at \$11.651 per barrel, with a market price of \$8.322 per barrel (the posted price is set at 1.4 times the market price). At this posted price, royalty payments to the host governments will be \$1.46 per barrel (12½ percent of posted price). Assuming production costs of \$0.12 per barrel, income taxes are 55 percent of \$10.08 (posted price minus royalties and costs), or \$5.54 per barrel. Thus, total government revenue (royalties plus income tax) is \$7.00 per barrel, or 84 percent of the market price. For oil sold directly by the government as a result of its participation share, revenue per barrel is estimated at approximately \$8.00.

Estimates of the behavior of petroleum prices over the 1973–80 period are tentative because they are subject to many variables, particularly noneconomic factors. The relative bargaining strength and attitudes of OPEC countries, energy and import policies of industrialized countries, and the political situation in the Middle East all play major roles in determining prices. Therefore a broad range of estimates for the Persian Gulf price in 1980 must be utilized.

Adelman, forecasting increases in income taxes and royalty payments before the 1973 increases, predicted a price of $5.00 per barrel in 1980.[8] The *Oil Daily* reported in July 1973 that experts in Iran predicted a price of $10.00 per barrel for Iranian crude in less than a decade if energy substitutes are not developed.[9] James Akins thought $10.00 per barrel would not be an unreasonable expectation.[10] A complication is that prices might peak in the nearer term and then decline by 1980, because the price elasticity of demand is likely to be higher over the longer run, and by 1980 alternative sources of supply are likely to emerge.

Revenues

This study accepts a level of government revenues in 1980 consistent with an estimate of $7.00 per barrel and an output of 16 million b/d. There is likely to be some tradeoff between price and volume, given Saudi Arabia's predominance in world petroleum markets and the common policies of the OPEC countries. Thus, the government revenue estimate above would still hold if there were higher prices but lower volumes of output (and vice versa). For example, if the 1980 revenues were $5.00 per barrel and output were 22 million b/d, government revenues would be almost equal to those generated by $7.00 per barrel at a production level of 16 million b/d.

Based on these assumptions, Saudi government revenues would approximate $41 billion for 1980. Cumulative revenues for the 1973–80 period would approach $233 billion. Given a population of approximately 5 million, revenues per capita would be almost $8,200 by 1980; this compares with approximately $1,000 per capita in 1973.

3

Government Revenues and Expenditures: The 1960s Experience

Revenues

In the decade between 1960 and 1970 government revenues increased fourfold, from $350.9 million to $1,385.8 million (see table 1). Then in the two years 1971 and 1972 they rose substantially—by 50 percent in 1971 (to $2,072.5 million) and another 41 percent in 1972 (to $2,933.6 million). In riyals, the increase is less, since in 1972 the riyal was revalued from $1 = 4.500 riyals to $1 = 4.145 riyals.

Not only did revenues increase rapidly in the 1960s, but they increased at a considerably irregular rate; for example, they increased 8.9 percent in 1961, 21 percent in 1962 and over 25 percent in 1966 and 1967, but decreased slightly in 1960 and 1968.

Appropriations and Expenditures

In order to have some basis for forecasting the expenditures of the Saudi government over the 1973–80 period, we will examine expenditures during the 1960–72 period. To the extent possible the allocation of resources between consumption and development purposes will be examined and explained.

The breakdown of Saudi expenditures into consumption and development expenditures is somewhat arbitrary. The Saudi budget category "Development" includes a wide range of expendi-

TABLE 1. Revenues, Annual Rates of Change: 1959–72

(millions of dollars)

Year[a]	From oil[b] (1)	From other sources[c] (2)	Oil as % of total (3)[d]	Total (4)	Percentage change[e] in revenue (5)
1959	313.1	49.3	86	362.4	
1960	333.7	17.2	95	350.9	−3.2[f]
1961	377.6	4.6	99	382.2	8.9
1962	409.7	53.6	88	463.3	21.2
1963	607.7[g]	70.4	90	678.1	13.4
1964	523.2	67.0	89	590.2	12.3
1965	662.6[h]	68.3	91	730.9	16.1
1966	787.9[i]	112.6	88	900.5	27.5
1967	909.1[j]	236.9	80	1,146.0	27.9
1968	926.8	170.3	84	1,097.1	−1.8
1969	949.0	281.0	77	1,230.0	12.1
1970	1,214.0	171.8	88	1,385.8	12.7
1971	1,884.9	187.6	91	2,072.5	50.0
1972	2,734.1[k]	199.5	93	2,933.6	41.6

Sources: Al Bashir, SAMA (1972).

[a]Corresponding to the Saudi fiscal year.

[b]Royalty and income tax payments.

[c]Income tax on domestic individuals and business, customs, duties, etc.

[d]Column 3 represents column 1 as a percentage of column 4.

[e]Based on revenues that do not include special payments noted in column 1.

[f]In terms of riyals, government revenues increased because of the devaluation of the riyal from \$1 = 3.75 riyals to \$1 = 4.50 riyals.

[g]Includes special payment of \$152.5 million.

[h]Includes special payment of \$46.0 million.

[i]Excludes \$45.2 million value of oil deliveries to the Saudi government resulting from the participation agreement.

[j]Includes special payment of \$29.4 million.

[k]Includes special payment of \$29.3 million.

ture uses but basically corresponds to a projects budget. The government subsumes all construction, including that of government buildings and mosques, under "development." In this study we accept the categories used by the Saudi government. The result is that the data probably understate the proportions of government spending that represent consumption rather than development, as these terms are typically utilized. On the other hand, administrative and operating expenses of ministries which play an important development function, such as the Ministry of Transportation and Communication, are listed in the consumption category. If there were more detail concerning the functions of consumption expenditures a portion could be assigned to development, but such detail is lacking. To add to the problem, some defense expenditures are likely to be hidden in other budget categories.

Appropriations

Government appropriations to the consumption category increased approximately two and one-half times from 1959 to 1970, from $304 million to $730 million (see table 2). This rate of increase was considerably less than the rate of increase of revenues over the same period. Development appropriations, as measured by the projects budget, increased from $15 million in 1959 to $596 million in 1970. Thus, the growth of development appropriations has been considerably faster than appropriations for consumption purposes. Development appropriations as a percent of total appropriations increased from 5 percent in 1959 to 20 percent in 1964 to 45 percent in 1970.

As compared with appropriations, government expenditures in the 1964–72 period increased more slowly; only for 1965 and 1966 were they greater than appropriations (see table 2). For the period from 1964 through 1970 consumption expenditures either exceeded or matched consumption appropriations. But for each year of the entire 1964–71 period, development expenditures were less than development appropriations. It is evident from this record that the government was not capable of implementing development projects and maintaining the rate of construction implied by development appropriations, especially in the mid-1960s. In addition to administrative delays and the lack of certain skills, there were the typical delays in receiving bids, letting contracts, and implementing the contracts. Such delays are inevitably longer when contracts are let to foreign firms, which was true of a considerable proportion of the major development projects, and particularly those in transportation and communication. After 1967 implementation improved, so that by 1971 development expenditures had increased to 88 percent of appropriations. In 1972, however, because development appropriations were more than doubled from the preceding year, expenditures declined to 74 percent of appropriations.

The data indicate that in any given fiscal year the government transferred funds appropriated but not spent for development to other uses. A considerable proportion of the transferred funds was probably used for government salaries and other administrative costs, and for national defense and internal security, although

TABLE 2. Revenues and Appropriations 1959–73, with Patterns of Expenditure 1964–72[a]

(millions of dollars)

		Consumption			Development			Total		
Year	Revenues	Appropriations	Expenditures	Expenditures as % of Appropriations	Appropriations	Expenditures	Expenditures as % of Appropriations	Appropriations	Expenditures	Expenditures as % of Appropriations
1959	362.4	304.0			14.7			318.7[b]		
1960	350.9	274.2			24.4			298.6		
1961	382.2	267.6			64.7			332.3		
1962	463.3	324.4			88.9			413.3		
1963	525.6	420.0			105.6[c]			525.6		
1964	590.2	474.7	475.3	100	122.2	58.4	48	596.9	533.7	89
1965	684.9	522.2	672.4	129	169.3	80.7	48	691.5	753.1	109
1966	872.9	568.7	798.0	140	311.6	163.1	52	880.3	961.1	109
1967	1,116.7	735.1	840.4	114	381.6	175.1	46	1,116.7	1,015.5	91
1968	1,097.1	620.2	675.6	109	476.9	378.7	79	1,097.1	1,054.3	96
1969	1,230.0	658.9	684.4	104	571.1	413.3	72	1,230.0	1,097.7	89
1970	1,385.8	729.8	723.1	99	596.0	480.7	81	1,325.8	1,203.8	91
1971	2,072.5	840.9	774.0	92	576.9	505.1	88	1,417.8	1,279.1	90
1972	2,933.6	1,325.9	1,226.5	93	1,215.0	905.6	74	2,601.2[d]	2,132.2	82
1973[e]	5,190.0	1,737.8			1,801.1			3,538.9[f]		

Sources: Al Bashir, SAMA (1960, 1966, 1969, 1971, 1972).
[a]Figures on Expenditures 1959–1963 were unobtainable.
[b]Excludes repayment of government debt.
[c]Average of 1962 and 1964 development appropriations.
[d]60.3 million dollars allocated to budget reserve.
[e]Estimated. Appropriations based on fiscal year ending August 1, 1973.
[f]205.9 million dollars allocated to budget reserve and saving.

details of these expenditures are not published. For the two years in which actual expenditures were greater than appropriations —1965 and 1966—some expenditures were probably financed out of special payments received from Aramco as a result of negotiated settlements.

Three major characteristics of the 1967–71 period are important. First, expenditures have approximated 90 percent of appropriations. Second, expenditures as a percentage of development appropriations have increased, averaging about 80 percent for the 1968–71 period and rising to 88 percent in 1971. Third, consumption expenditures were almost equal to consumption appropriations. Thus, more recently appropriations have become a much better indicator of actual expenditures than they were in the earlier period. That the Saudis have financed total current expenditures out of current revenues shows that budgetary control and performance in implementing development projects have improved.

In this respect, it is useful to review the concept of a balanced budget as viewed from the perspective of Saudi conventions. Since the financial crisis which gripped Saudi Arabia in the late 1950s, the government has emphasized balanced budgets. However, a balanced budget does not signify that current fiscal year expenditures equal current fiscal year receipts. By a balanced budget the Saudis mean that expenditures are not financed by issuing government debt. If the Saudi government finances part of current expenditures by drawing down reserves built up by budget surpluses in earlier years, this is regarded as a balanced budget. So is the situation in which excess revenues are allocated to external reserves. Such reserves have accumulated in the past primarily because cash expenditures on development projects have been less than appropriated expenditures. Budget policy in Saudi Arabia over most of the 1960s was to equate appropriations with anticipated revenues. A principal exception was the period from 1959 to 1962, when national debt was being retired. In March 1970, gold, foreign exchange holdings and investments of the Saudi Arabian Monetary Agency were valued at about $779 million, compared with approximately $641 million in April 1965.[12] Thus, most of the revenues during this five-year period were appropriated and spent.

Beginning in 1971 the situation changed. The amounts appropriated for expenditures were established at about 98 percent of anticipated revenues, with about 2 percent allocated to reserves ($36 million in 1971 and $60 million in 1972). However, it has been

TABLE 3. Revenues, Appropriations, and Expenditures: Annual Rates of Change, 1960–72

(percentage rate of change)

		Appropriations		Expenditures	
Year	Revenues	Consumption	Development	Consumption	Development
1960	−3.2	8.3	100.0	n.a.	n.a.
1961	8.9	−1.6	164.5	n.a.	n.a.
1962	21.2	21.3	37.1	n.a.	n.a.
1963	13.4	29.4	18.8	n.a.	n.a.
1964	12.3	13.0	15.8	n.a.	n.a.
1965	16.1	10.0	38.5	41.5	38.2
1966	27.5	8.9	84.1	18.7	102.1
1967	27.9	29.3	22.5	5.3	7.4
1968	−1.8	−15.6	25.0	−19.6	116.3
1969	12.1	6.2	19.7	1.3	9.1
1970	12.7	10.8	4.4	5.7	16.3
1971	49.6	15.2	−3.2	7.0	5.1
1972	41.6	45.2	94.0	58.5	79.3

Source: Table 2.

estimated that external reserves of Saudi Arabia in March 1973 were approximately $2.9 billion, an amount slightly more than total government appropriations for 1972.[13] Reserves were $779 million in March 1970 and the increase to $2.9 billion in March 1973 resulted from revenues higher than anticipated. Revenues rose from $1,385.8 million in 1970 to $2,072.5 million in 1971 and then to $2,933.6 in 1972, and appropriations and expenditures lagged behind. In 1971 appropriations were approximately two-thirds of revenues; in 1972, the comparable figure was 87 percent. Appropriations had been increased only 8 percent in 1971, reflecting the slow growth in revenues which characterized the 1967–70 period (see table 3). However, after revenues rose by 68 percent in 1971, the government increased appropriations by 70 percent in 1972.[14] The principal reason appropriations were only two-thirds of revenues in 1971 was that the latter were substantially underestimated. The SAMA Report for 1971 estimated revenues at $1,417.8 million, which was $654.2 million less than realized revenues.[15] The very large increase in appropriations for 1972, 70 percent, resulted from a better estimate of revenues and is characterized primarily by the almost doubling of the appropriations for development.

In preparing the 1972 budget the Saudis attempted to avoid another surplus like the one that resulted in 1971 from unanticipated revenue increases. So they matched expenditures to revenues and budgeted \$2.6 billion for each. But again, the realized revenues were higher than anticipated.

The budget figures for 1971 and 1972 are not directly comparable to earlier years. After the 1967 Arab–Israeli war, Saudi Arabia undertook aid programs to other Arab countries in an amount averaging approximately \$150 million per year. Prior to 1971 this aid was handled as an "out-of-the-budget" expenditure. The 1971 budget included \$88.9 million of this amount as a budgeted item, and for 1972 the full amount was budgeted as a subsidy. In other words, the budget estimates for 1968–70 understate appropriated amounts, but since we lack details of the composition of these accounts, we cannot determine by how much.

In the period for which data are available, appropriations for consumption increased more consistently year by year than appropriations for development (see table 3). In an attempt to explain the behavior of consumption appropriations for the 1960–70 period, al-Bashir estimated:[16]

$$C_g = -21.877 + \underset{(.054)}{.242\, G_R} + \underset{(.109)}{.634\, C_g} - 1 \qquad R^2 = 0.99$$

$$C_g = -61.344 + \underset{(.083)}{.294\, G_R} + \underset{(.089)}{.278\, G_R} - 1 \qquad R^2 = 0.99$$

where

C_g = government appropriations for consumption
G_R = government revenues

Al-Bashir concluded that in a given fiscal year the level of both revenues and consumption appropriations in the previous year were strong elements in explaining consumption appropriations. The previous level of consumption appropriations is important, he explained, because "when new expenditures are introduced in period t, the probability of these new expenditures being continued in period $t + 1$ is high."[17] Previous year revenues are important because often parts of these revenues are transferred into a reserve fund where they can be drawn upon in subsequent years.[18] Also, the previous year's experience helps to determine revenue estimates. The data show that unanticipated large increases in revenues have been appropriated in subsequent years, with the record for 1971–72 not being complete at this writing. Based on his analysis, al-Bashir concluded that the government's long-run marginal propensity to consume, as represented by consumption appropriations relative to revenues, is .661;[19] thus, approximately two-thirds of increases in revenue were used for consumption.

The major conclusions to be drawn from this analysis are these:

1. Appropriations have tended to rise proportionate to increases in revenues.

2. Expenditures for development have been less than appropriations, and often unspent allocations have been transferred to consumption appropriations and expenditures.

3. Expenditures for development have risen as a percentage of appropriated amounts.

4. Revenues have fluctuated considerably over this period, and this helps to explain why expenditures often diverge considerably from appropriations.

5. Appropriations for consumption have increased steadily, the increases averaging about 16 percent per year over the 1960–71 period, but consumption expenditures have fluctuated more than appropriations.

In addition to these general observations and before attempting to estimate revenues and expenditures over the 1973–80 period, an examination of the most recent budget experience is helpful for judging future policy. The information is fragmentary but reveals some trends and policies.

Consumption Expenditures

In the consumption category, it appears that appropriations for national defense increased from $275 million in 1972 to $900 million in 1973, and budgeted appropriations for defense were $1.6 billion for 1974. These appropriation figures predate the Arab–Israeli war, and it seems likely that Saudi defense expenditures will be even higher than the appropriated amounts.

Government spending for health and educational services is also increasingly rapidly. And the government, as the largest employer in the kingdom, has been absorbing an increased proportion of the labor force as better-educated Saudis find it difficult to get work in the private sector; this has become necessary because the petroleum sector employs fewer persons today than it did over a decade ago, and many major development projects are capital-intensive and thus offer limited employment opportunities.

Finally, there are some indications that the government has begun new efforts to improve consumption standards. One is that electricity, which has been quite expensive in Saudi Arabia, is now subsidized. Beginning in 1972 electric rates were reduced by 30 to 50 percent. Private power companies are guaranteed an annual rate of return of not less than 7 percent of paid-up capital if the lower rates cause reductions in profit margins below this level.[29] Significantly, the first phase of a five-stage program will introduce electricity to eighty small towns and villages, with the government supplying most of the capital. The introduction of electricity in such communities will of course change patterns of communication and consumption that have not changed for centuries. The second indication is that tariffs have been reduced twice since the beginning of 1972. Tariffs on items such as meat, tires, wood, cars, trucks, bicycles, air conditioners, and cement have been substantially reduced. Since customs duties are no longer needed to provide government income, and there is no operational balance-of-payments restraint, most tariffs are likely to be eliminated, and the list of imported items subsidized by the government will be increased.

These examples establish that the limits to government consumption spending are not obvious. In many areas, particularly national defense, this spending is rising rapidly. With no foreign exchange restraint, expenditures can be converted into real resources easily. Commercial and trade activity at the wholesale and retail levels may become the leading element of the private sector of the economy.

Development Expenditures

Development expenditures jumped by 94 percent in 1972. Over the 1964–71 period they fluctuated more than appropriations. The development budget suffered the most from the slow growth of revenues following the 1967 Arab–Israeli conflict, and from the commitments of aid to other Arab countries. With rapid and sustained increases in revenues predicted from 1972 on, there should be more willingness to commit resources to development projects.

It is unlikely that the rapid increases in revenues have radically altered development strategy; instead the traditional areas of focus will probably receive relatively more resources, with new emphases on housing and municipal services, petrochemicals, and desalinization projects.

Housing in Saudi Arabia has become relatively high priced. Land values in the major cities have risen rapidly; rent for an apartment suitable for a family of four in Riyadh was more than $100 per month in late 1973. Housing construction has been primarily a private sector activity. Government appropriations for housing have been relatively low, less that one-half of 1 percent of total appropriations. It was estimated that during the mid-1960s fewer "modern" new housing units were constructed than new family units were formed. Total housing requirements meeting minimal standards would be difficult to estimate, given the lack of statistics and the difficulty of establishing requirements for the one-fourth to one-third of the population that is nomadic. But to remedy the present excess of sub-standard housing in the major urban centers of the east and west coasts and Riyadh requires a decade of major building programs. Housing remains the neglected sector of social overhead capital and one that is an obvious target in any program aimed at improving the country's standard of living. It is one in which complementary inputs are more easily obtained than they are for such areas as health and education. If the government were to subsidize housing construction and rentals, this would result in a major increase in expenditures the government classifies as development projects.

Such a commitment of resources would be consistent, too, with a growing interest and emphasis on municipalities and urban planning in the kingdom. In Saudi Arabia municipalities have no independent sources of revenue and are dependent on allocations by the national government. The central authorities are responsible for all the costs associated with improving settled places, including water and sewage systems, transportation, other municipal services, efforts at urban renewal, and beautification projects. In the larger cities (Riyadh, Jeddah, Dammon), fairly rapid urbanization has brought about the social costs usually associated with this process. As elsewhere, any attempts to cope with these costs will involve a very large commitment of resources.

Development expenditures are more closely tied to the availability of other local inputs, planning and administration, etc., than are consumption expenditures. Development appropriations experienced more variation during the 1960s than did consumption appropriations and also increased more rapidly, at least until 1969. But with a lower rate of revenue increase during the late 1960s, development appropriations were curtailed and were only $576 million in 1971 compared with $571.1 million in 1969. Compared with the 1964–67 period, when expenditures were only about one-half of appropriations, the proportion for the 1968–71 period was about 80 percent.

In 1972, in response to the large 1971–72 increase in revenues, development appropriations were more than double the 1971 level. It was reported that government development expenditures were only about 60 percent of allocations during the 1972–73 period, which indicates that the implementation of projects had fallen considerably below the performance of recent years.[30] The budgets indicate that the Saudis intend to increase the proportion of resources being utilized for development purposes, and that allocations will increase rapidly as revenues increase. However, development expenditures as a proportion of allocations will undoubtedly be lower than the 1968–71 period, although expenditures on development projects increased by approximately 50 percent in 1972 compared with the 1971 level. This demonstrates that the Saudis have increased their capability for implementing development plans.[31]

Development activity can increase extremely rapidly in some areas. For example, in 1972 consumption of cement rose by 25 percent, imports of machinery increased by approximately 23 percent, and the generation of electricity rose by over 25 percent.[32] Project allocations to the Ministries of Transport and Communication and Interior (for municipalities) were more than doubled in the 1972–73 period; appropriations for education increased tenfold over the same period. Those who argue that the Saudis cannot increase their spending very rapidly do not seem to recognize the recent Saudi experience. The abilities of Saudi authorities to organize and plan project and other expenditures have improved considerably during the past five years.

4

Revenues and Expenditures, 1973–80

Revenues

It can be argued that the projected magnitudes and rates of change of revenues for 1973–80 are sufficiently different for these years that past experience can provide little guidance in anticipating Saudi propensities to spend and to save. Revenues of $40 billion by 1980, compared with revenues of $3 billion in 1972, represent such an increase in wealth that entirely new relationships may be established between revenues and spending for consumption and development. Any quantitative relationships established for the 1960s must be qualified by considerations such as these when constructing a forecast. However, the experience of the 1960s is helpful in making judgments about the government policies that will play a decisive role in determining how the revenues will be disposed of. If the royal family maintains its primacy and the line of succession to King Faisal continues as presently constituted, one can reasonably assume that the basic attitudes and policies of government leadership will carry over into the period up to 1980. There is no indication that the country has potential political leadership that would differ radically from that of King Faisal's regime. In this respect, one must be careful to differentiate between policy decisions related to the Middle East conflict and more fundamental long-run attitudes toward the utilization of resources.

It is also important to keep in perspective values projected seven, ten, or fifteen years ahead. In magnitude, 1980 earnings compared with those of 1973 should be unprecedented, but the rates of change may be no more rapid than those in the early 1950s. It is understandable that some question Saudi Arabia's ability to absorb a substantial proportion of these income increases. But answers can be too quickly given. Would one have asked the same questions about a trillion-dollar GNP level for the United States from the vantage point of a national output of $284 billion in 1950?

It is, of course, the rapid increase in and the magnitudes of revenues which will most influence the spending response of the Saudi government. In addition to the rate at which revenues increase, however, one must consider the predictability of the rates of increase. The predictability of revenues will influence development strategy and the planning horizon for expenditures.

Revenues in 1971 were 50 percent higher than in 1970; and in 1972 almost 42 percent higher than in 1971. Revenues of approximately $5 billion in 1973 is an increase of almost 75 percent over 1972. In 1974 revenues are likely to increase almost 400 percent, but the rates of growth of revenues should be small in subsequent years. For the 1974–80 period, revenue is expected to increase in large dollar amounts, but the rate of increase is expected to moderate. If we look at the rates of increase associated with 1980 $7.00 per barrel oil revenues and 16 million b/d production, it is clear that the very large dollar increases in revenues are associated with declining rates of increase (see table 4). Of course, production and/or price changes could occur more sporadically, and rates of increase of revenues for any one year could be appreciably higher, or lower, than the range specified. However, the average

TABLE 4. Revenues, Projected Annual Rates of Change, 1974–80

Year	Percent change
1974	388
1975	11
1976	10
1977	9
1978	9
1979	8
1980	7

NOTE: Assumes revenues of $7/barrel and production increasing steadily to 16 million b/d by 1980.

rate of growth during the 1973–80 period would approximate the levels shown.

The substantial variations in the rate of growth of revenues in the 1960s hampered efforts of the Saudi government to program their use. Especially with the background of the financial chaos of the late 1950s, a period when revenues were squandered and which led to the replacement of King Saud by King Faisal, the government was understandably hesitant to commit itself to programs if it was uncertain that they could be financially supported in the future. While the government can not be sure of the year to year timing of revenues because of some uncertainty about production and prices, the general magnitude of revenues is much more predictable for the 1973–80 period than the decade of the 1960s. Any substantial shortfall within the predicted range is likely to be self imposed. The forecasts of revenues allow for major commitments of resources to a wide range of programs, whether these be development projects or efforts to raise consumption standards more directly through increased imports. Also as the Saudi government accumulates reserves, the importance of having accurate forecasts of revenues is less important than when expenditures must be financed out of current revenues. Essentially, planners will be free of any financial constraint, either in terms of foreign exchange or the riyal, in the establishment of expenditure programs if considerable extravagance is avoided.

Expenditures

Because the rate of increase of oil revenues will be considerably higher in the 1973–80 period than in the previous decade, and because the absolute amounts are so great, it is not possible to project ahead mechanically the expenditure patterns just analyzed. Expenditures have tended to follow revenues quite closely even in years when revenues have risen substantially, and even when revenue changes have been unpredictable. As viewed by Saudi policy makers, there is no evidence that desired expenditures have been less than revenues. On the contrary, during a period when administrative skills and budgetary procedures were in a formative stage, expenditures rose rapidly. True, development expenditures did not keep pace with allocations, but these funds were transferred to other uses when this occurred.

The concept of absorption capacity, well entrenched in the development literature, also bears on a judgement of the extent to which government spending will accompany rises in petroleum revenues. This concept recognizes limits to the ability of a less-developed country to make efficient use of financial resources and new capital equipment. The limits are related to inadequacies of complementary inputs, primarily skilled labor and managerial and entrepreneurial talent. In this sense, absorptive capacity is an important factor in analyzing Saudi Arabia's ability to utilize efficiently its substantial revenue increases. Complementary inputs are relatively scarce, and aside from relatively capital-intensive investments in petrochemicals and similar ventures, the opportunities for investment in manufacturing and to some extent agriculture are limited. Saudi Arabia's capacity to absorb real capital falls considerably short of the availability of financial capital. As expressed in a Petroleum Press Service editorial of early 1973:

> The rate at which a predominantly agricultural community—still more, a pastoral community—can become industrialized is certainly not governed only by the amount of capital available for investment. Inadequate education, lack of technical know-how, social customs, religious beliefs—all these may limit the rate of economic growth. And the process of economic and social change cannot be accelerated beyond a certain point without disrupting the fabric of society and inflicting intolerable hardship on the people concerned. In some OPEC countries, lack of capital will soon cease to be the limiting factor in economic development.[11]

In describing the situation for Saudi Arabia, analysts may exaggerate the inability to absorb capital adequately as a determining factor in limiting government expenditures. Such a viewpoint neglects the potential role of government as a consumer and as a subsidizer of consumption. With no foreign exchange constraint, and with major increases in consumption being largely identified with imports, the major constraint to government spending is its willingness to allow consumption standards to rise. The same reasoning applies to national defense expenditures. By recognizing that "expenditures" absorptive capacity differs considerably from capital absorptive capacity, one can reach quite different conclusions. The relative weights assigned to investment as against consumption in government expenditure decisions and the composition of expenditures within each of these categories is crucial to determining the effects of absorptive capacity on government expenditures.

The strategy of economic development in Saudi Arabia has been a fairly traditional one. Expenditures have been concentrated on social overhead investment in transportation, communications, and education. The government has invested in agriculture in the form of water and irrigation projects and efforts to provide more permanent settlements for the Bedouin. The government has taken an active role in establishing industries in such areas as petrochemicals and steel. However, even though government petroleum revenues account for such a substantial proportion of national income (approximately half), the government has interfered very little in the private sector of the economy. Direct subsidization has been restricted primarily to assistance in providing seeds and machinery to agriculture, and import subsidies for drugs and medicine and such foodstuffs as food grains (wheat and rice), sheep, vegetable oil, and milk. This is substantially different from policies in Kuwait, for example, which have emphasized social welfare projects, subsidized prices for agricultural output, provision of land, and the like. Of course, Kuwait's wealth per capita has been considerably greater over the past decade than Saudi Arabia's.

How the Saudi authorities view appropriate budget and development strategy will be an important determinant of the proportions of revenues that will be spent and saved. In other words, while during the 1960s government spending was essentially determined and limited by the level of government revenues, the expected magnitude of revenues in the 1973–80 period gives much more discretion to the Saudi authorities and more importance to budget and development strategy.

An important element of this strategy will be the extent to which the Saudi government will directly subsidize current private consumption out of petroleum revenues. Essentially, it would appear to have two major options. The first is to continue the policy of rapid expansion of government expenditures which characterized the 1960s, concentrating on social overhead investment, productive investment in agriculture, the petrochemical industry, other selected heavy industries such as steel, governmental operating expenses of the existing ministries, and national defense. The second option would add to the above the subsidization of private consumption, primarily through expenditures on imports of consumer goods and subsidization of housing. The second option will result in a faster growth of expenditures, of course.

Factors Affecting Expenditures

Five major factors might play a substantial role in shaping decisions about future levels and patterns of expenditures. First, the Saudis are understandably concerned that too rapid increases in wealth and changes in the standard of living will undermine Saudi traditions of culture and religion. The traditions and religion of the Saudi people have always been important to King Faisal; he has attempted to limit the impact of Westerners and other foreigners over the past decade. He has discouraged tourism for this reason. Similarly, "It is also reported that the Saudi government . . . is not prepared to accept the degree of imported labor which has made the populations of Dubai and Abu Dhabi, each between 75,000 and 90,000, three-quarters immigrant."[20] The same source quotes Prince Saud el Faisal, son of the king and deputy petroleum minister, as being concerned with "how much we can produce without harming ourselves."[21]

In assessing these concerns, we must take into account the rate and degree of change the country has already experienced. Petroleum revenues were $56.7 million in 1950, $333.7 million in 1960, and $1,154 million in 1970. Both the rates and sizes of increase are considerably smaller than what will occur during the 1970s, but it can be argued that Saudi Arabia was considerably more isolated and far less ready to absorb the income and to adjust to non-Saudi ideas and customs than it is today. To a considerable extent, the Western intrusion into Saudi Arabia has already occurred. For example, there has been a significant expenditure on education. It seems unlikely that this commitment could be reversed; the growing number of educated Saudis results in a rising proportion of the population looking for new jobs and with expectations of higher incomes. It will require astute economic and political management to accommodate religion and traditions to these desires. King Faisal has directed efforts toward introducing the population to the changing circumstances in a manner most consistent with Saudi traditions and religion rather than attempting to perpetuate the status quo.

Another factor affecting future expenditures will be the government's concern for the reactions of other less-developed countries, particularly other Arab states, to Saudi Arabia's rising income and wealth. Given the scarcity of resources in some of these countries, and what is expressed as a common cause in Arab–Israeli disputes, an accumulation of revenues and wealth by Saudi Arabia could lead to pressures for it to make loans and grants to its poorer Arab neighbors. Since 1968 Saudi Arabia has made annual grants averaging $147 million to other Arab countries. It was reported that Saudi Arabia intended to compensate Egypt if it loses its $87.5 million annual subsidy from Libya as a result of disagreements between those two countries.[22] Prior to the 1973 Middle East war, King Faisal reportedly was considering buying military planes for Egypt with the intent of reducing Soviet influence there.[23] Furthermore, Saudi Arabia has helped finance war expenditures of the Arab belligerents in the war with Israel. And, over the longer run, Saudi Arabia wants to exercise leadership in the entire Moslem world; this suggests potential aid to such countries as Pakistan. As was suggested by James Akins, grants to other Arab and Moslem countries "would be in perfect consonance with Islamic law practice, which demands twice as much *zakat* from income derived from lands fed by God-given rain as from lands irrigated by man."[24] Pressures will undoubtedly build for Saudi Arabia to aid other less-developed countries, especially those which are predominantly Moslem, whose balance of payments and economies are hurt by the substantial increases in petroleum prices. Saudi Arabia's foreign aid expenditures are likely to increase substantially as its income and wealth grow.

A third factor affecting expenditures is that pressures will continue to be applied for the government to share its income and wealth with all elements of the Saudi population. There seem to be some misconceptions about the degree of freedom Saudi authorities have in the formation of policy as it relates to the desires of the general population. Saudi Arabia's political system is an absolute monarchy. This leads to such observations as "Political activity is virtually nil"[25] and "King Faisal of Saudi Arabia shows no interest in political institutions."[26] If these comments mean that Saudi Arabia has few of the political institutions of Western political democracy, then they are accurate. But if they mean that the king does not have to respond to the mix of interests and aspirations of such groups as the extended royal family, Bedouins, religious groups, a growing educated class, and large numbers of civil servants, they distort the picture of Saudi political life. To a considerable extent government policies in Saudi Arabia emerge from a mix of diverse interests and conflicting goals similar to those in any of the Western democracies. With the relatively small population and its tribal base, there is a strong personal element in the conduct and formation of government policy. It is difficult for the government leadership to ignore pressures by the various elements of the population who want to share in the growing wealth by achieving rising standards of living.

Another factor affecting expenditures is the question of the appropriateness, in terms of expenditure policy, of using current revenues for current consumption instead of for productive purposes. Petroleum is an exhaustible resource, and there is understandable concern for diversifying and broadening the base for income-generating activities. In the case of Saudi Arabia, expenditure strategy is conditioned by some special features. First, government income outruns the opportunities for productive investment, given the resource base and the levels of skills of the population. While obvious potential for investment in petrochemicals and other industries exists where natural gas and petroleum can be utilized, the implementation of projects in this sector will be slow, and, being capital-intensive, such projects will not offer widespread employment opportunities. It remains to be demonstrated that a large number of directly productive and economically feasible investment projects can be generated during the balance of the decade. Investments are likely to remain concentrated in social overhead. Second, the proven and potential petroleum reserves of Saudi Arabia are so substantial that diversification may proceed more slowly as complementary inputs, primarily administrative and labor skills, are developed. Diversification will be more feasible ten years from now than it is today. Third, petroleum is a national resource, and the income generated by this sector is approximately half of total income. That this income accrues directly to the government should not obscure the appropriateness of committing a portion of these resources for consumption purposes. It seems likely that the relative proportions of government revenues directed toward consumption rather than for investment will be based on the degree to which resources can be invested productively and the relative weight attached to raising consumption standards.

Finally, it seems likely that increased revenues will result in major increases in defense spending. The ill-defined nature of national security makes it difficult to establish limits to defense expenditures, given the rapid rise in government income. There is some evidence that unspent development allocations in the past have been used for national defense. Military and subversive activities of South Yemen on the Arabian peninsula and security concerns about the Arabian Gulf as it relates to Kuwait and other oil-producing states are likely to heighten in future years. Saudi Arabia's increasing wealth will in itself heighten concern by its authorities to strengthen its defense posture.

An early indication of the thrust of the Saudi defense effort was revealed by U. S. Assistant Secretary of State Joseph J. Sisco. In remarks before the House Near East Subcommittee he stated,

> To give some current examples, we are going forward with the Saudis with plans to help them develop a modest naval force of small ships suitable for the protection of their extensive coastlines; we are selling F-5 Freedom Fighters to replace older American aircraft; we are now initiating a sales program to reorganize and equip elements of the National Guard [a paramilitary internal security force]; and American companies are continuing to assist in improving Saudi Arabia's air defense capabilities. Recently, we have agreed in principle to consider the sale of a limited number of F-4 aircraft.[27]

More recently, the Saudis have had discussions with the French concerning purchases of military equipment, particularly aircraft, and the Saudi Arabia–United States agreement of June 1974 includes cooperation in building up Saudi military forces.

Because military hardware will be imported, and because judgments concerning the effectiveness with which the equipment can be used will be difficult to make, defense expenditures have risen and will undoubtedly continue to rise considerably. For 1973, published defense appropriations were approximately $900 million, or 27 percent of the national budget.[28] Undoubtedly they have risen since the budget was published. Even in the absence of a firm rationale for such expenditures, the inclination toward the limitless requirements of national security would result in major expenditures for these purposes.

Any forecast of annual expenditures is also dependent on estimates of year-by-year increases in revenues. For this study, the assumption used is that after the enormous increase in 1974, revenues will rise steadily to the estimated level of 1980. While changes in prices and output are likely to be much more erratic than those presumed in this study, these changes cannot be anticipated and any other assumptions would be arbitrary too. Thus, forecasts of annual revenues, expenditures, and savings should be regarded only as rough indicators of the magnitudes that might be expected.

Two major considerations underlie the estimates of expenditures:

First, for 1974–75, it is expected that Saudi Arabia will expend extraordinary amounts on national defense and assistance to other Arab countries.

Second, the Saudi Arabian government is likely to continue and accelerate its efforts to increase consumption standards in the kingdom. It seems likely that consumption-type expenditures will increase as a proportion of total government expenditures.

Based on these and other considerations discussed earlier in the paper, the forecast assumes that government expenditures will be 80 percent of revenues in 1973 and will increase by 50 percent for each of the following two years. Expenditures will increase substantially after 1975, but the rate of increase will decline. It seems reasonable to assume that the Saudis will be capable and desirous of expanding expenditures at a rate of 30 percent per year over the 1976–80 period; this rate of increase has been approximated in the past even when revenues were not as large or as predictable as they will be. While the rate of increase may be somewhat more or less than the assumed 30 percent, projections based on this rate will give a reasonable estimate of Saudi government expenditures over this period.

5

Savings

It is projected that by 1980 Saudi expenditures will be $33.4 billion with revenues estimated at $40.9 billion (see table 5). Based on the assumptions outlined above, Saudi Arabian savings will total $108.7 billion over the eight-year span from 1973 to 1980. Because expenditures are expected to increase rather steadily over the 1976–80 period, if revenues are higher than those projected in table 5, savings will increase correspondingly; similarly, lower rates of revenue increases will decrease savings. It seems likely that the rate of projected revenue increases would have to decrease considerably before the level of government expenditures would be substantially affected.

The Implications of Saudi Foreign Exchange Savings

If by 1980 Saudi Arabia were to hold approximately $110 billion of short- and long-term foreign assets, it obviously would emerge as a major participant in international financial markets. The Saudi Arabian government will likely be the largest single holder of international financial assets in 1980. While we can only conjecture about the Saudi government's behavior in this role, some observations may assist in assessing the impact of this new element in international financial markets.

TABLE 5. Revenues, Expenditures, and Savings, Projected, 1973–80

(billions of dollars)

Year	Revenues[a]	Expenditures[b]	Savings
1973	5.0	4.0	1.0
1974	24.4	6.0	18.4
1975	27.2	9.0	18.2
1976	29.9	11.7	18.2
1977	32.6	15.2	17.4
1978	35.4	19.8	15.6
1979	38.1	25.7	12.4
1980	40.9	33.4	7.5
Total	233.5	124.8	108.7

[a]It is assumed that production increases steadily from 9.57 million b/d in 1974 to 16 million b/d in 1980. Revenues are assumed to be $7 per barrel beginning in 1974. There has been no attempt to estimate revenues from any source other than crude petroleum.

[b]It is assumed that expenditures increase by 50 percent in 1974 and 1975 and by 30 percent in each subsequent year.

The Composition of External Assets

The Saudi government categorizes its external assets as gold, foreign exchange, and foreign investment. The difference between foreign exchange and foreign investment is not clear. "Most Saudi oil revenues are paid into accounts with Morgan Guaranty and Chase Manhattan. Apart from a small margin kept to meet current obligations, they are then placed through a number of banks and brokers into short-term holdings other than equities."[33] For purposes of analyzing the composition of external assets, we assume that except for small working balances the two categories, foreign exchange and foreign investment, are indistinguishable and measure investments in short-term assets.

Saudi external assets in gold, foreign exchange, and foreign investments increased from $170.64 million in 1960 to $1,349.33 million in 1971, $2,487.09 in 1972, and $3,321.72 in March 1973 (see table 6).

As a result of its experience during the late 1950s, when its foreign exchange reserves were depleted and imports had to be curtailed, the Saudi government has maintained high levels of foreign exchange and short-term assets to guard against adverse

TABLE 6. External Assets, 1960–73

(millions of dollars)

Year	Gold	Foreign Exchange	Foreign Investment	Total
1960	17.81	152.83	—	170.64
1961	65.37	167.91	—	233.28
1962	77.79	206.56	—	284.35
1963	77.72	426.40	—	504.12
1964	77.72	506.37	—	584.09
1965	77.56	630.89	26.22	734.67
1966	77.56	663.78	65.33	806.67
1967	77.56	661.11	158.44	897.11
1968	127.56	519.56	183.11	830.23
1969	127.56	464.44	154.89	746.89
1970	127.56	520.44	174.89	822.89
1971	127.56	1,039.33	182.44	1,349.33
1972	138.48	2,152.95	195.66	2,487.09
1973[a]	153.89	2,904.29	263.54	3,321.72

[a]March, 1973. Exchange rate is \$1 = 3.73 riyals.

effects from declines of petroleum revenues. Beginning in 1963, after a period when reserves were built up, total foreign assets have averaged about one-third the level of Saudi GNP, approximately one and one-half times the value of annual imports, and about equal to the annual value of petroleum revenues. The accumulated levels of foreign exchange savings projected for 1980 (see table 3) would raise total foreign exchange assets to two and one-half times the projected annual petroleum revenues. Conservative policies have been prudent in the past, given the heavy reliance on the petroleum export sector, which was subject to short-term disruptions and fluctuations. However, given the relative predictability of revenues in the near future, the high value of future foreign assets, and a growing reliance on tanker shipments directly out of the Arabian Gulf rather than through TAPLINE (Lebanon), Saudi authorities will have more flexibility in deciding the composition of foreign assets than they have had in the recent past.

Equity Investments

In assessing the utilization of foreign exchange savings, a new factor is the potential role of foreign equity investment by the

Saudi government. Minister of Petroleum Ahmad Zaki Yamani, has expressed the government's interest in investments and in the consuming countries' downstream oil operations.[34] A new ministerial committee, headed by Prince Fahad al-Saud, who is minister of the interior and the apparent successor to King Faisal, has been established to work out government foreign investment policy. It is reported that the government is ". . . considering diversification outside of the petroleum sector instead of further investments in the oil industry."[35]

There are good reasons for the Saudis to invest in other countries. Productive investment opportunities within the kingdom are smaller than the funds available, and foreign investment is a productive outlet for funds that cannot be utilized for domestic projects. Complete dependence on petroleum exports for foreign exchange earnings makes the Saudi economy vulnerable, and other sources of foreign exchange would ease the situation. Over the longer run, petroleum is an exhaustible resource, and alternative sources of foreign exchange earnings would enable the Saudis to prepare for a time when their dependence on petroleum should lessen. And income from other foreign investments would be helpful during any short-run disruptions in petroleum exports. During the 1973–80 period, when revenues are expected to increase so rapidly, Saudi Arabia is likely to have both the interest in and capability of diversifying its sources of foreign exchange earnings.

This investment may take any of three forms. One is typified by Iran's proposed joint venture with Ashland Oil Company. In this agreement Iran would have acquired a 50 percent share of Ashland's refining and marketing operations in New York State in return for a guaranteed long-term oil supply.[36] Saudi Arabia, in placing its participation share of crude, may decide to enter into similar arrangements. The National Petroleum Council has estimated that in the United States the cumulative capital requirements for oil pipelines, gas transportation, refining, and tankers and terminals will range from $59 billion to $98 billion (in constant 1970 dollars) over the 1971–85 period.[37] The requirements for all United States energy industries are considerably higher, of course. For the 1970–85 period world capital requirements for the petroleum industry were estimated at $565 billion by the Chase Manhattan Bank.[38] Saudi Arabia may decide to invest a portion of its assets in this sector.

A second type of equity investment would be portfolio investment through the major world capital markets. Saudi Arabia would acquire income-earning assets without any attempt to get a voice in managerial affairs, a diversified portfolio, and reduced dependence on petroleum activities, whether at home or abroad.

A third type of equity investment would be direct investment in nonpetroleum enterprises by getting 50 percent or more equity. These would become essentially Saudi national companies under Saudi managerial control.

For the period under consideration, it does not seem likely that Saudi Arabia will attempt much investment of the third type; the available managerial talent will likely be concentrated on domestic activities. But there may be considerable equity investment of the first two types. It would seem to be in the long-term interests of Saudi Arabia to diversify its sources of foreign exchange earnings, and it would not be surprising to see a considerable proportion, perhaps as much as half, of foreign exchange savings used in this manner.

Short-term Assets

In this study foreign exchange savings for some years are projected to be as high as $18 billion; the cumulative levels of savings in 1980 are almost $110 billion. Some portion of these funds is likely to have been appropriated but not spent, and because Saudi Arabia must be concerned with its international liquidity position to protect itself against short-run disruptions or fluctuations in petroleum revenues, considerable amounts may be invested in short-term assets or held as working balances. Short-term assets might be as much as one-third of total assets, or approximately $35 billion. By way of comparison, banks in the United States reported liabilities of over $16 billion to Japan in late 1972, and $12 billion to Germany in May 1973.[39] Saudi Arabia, under these conditions, would emerge as one of the major holders of short-term dollar assets.

While Saudi holdings of international short-term assets are difficult to identify from available statistics, it appears that most are held in the form of Euro-dollar deposits. This has apparently resulted from the relatively higher interest rates on Euro-dollar deposits and the fact that almost half of Saudi petroleum exports go to Western Europe. It seems reasonable to expect that short-term asset holdings will be concentrated geographically in areas that are the principal importers of Saudi oil, although increasing amounts will be placed in the United States. Saudi Arabia may be expected to increase pressures on Japan to make its financial markets more accessible.

The International Exchange Rate System

With the emergence of Saudi Arabia and other petroleum producers as major capital-exporting countries, a strong case can be made for retaining exchange rate flexibility as a feature of the international exchange rate system. It is fruitless to try to judge whether or not Saudi Arabia's actions as a capital exporter will be stabilizing or destabilizing in international financial markets, basically because it is not possible to guage future intentions. We might expect, in any event, major adjustments to reflect the emerging international financial positions of Saudi Arabia and others. Flexibility of exchange rates should make the transition easier. And if in the future Saudi Arabia behaves in such a manner as to create conditions of instability, flexible exchange rates will tend to penalize such action. Substantial and rapid movements of funds will be accomplished only by incurring capital losses. On the other hand, fixed exchange rates tend to reward such movements of funds. Flexible exchange rates will assist in the period when new trade and financial relationships are being established; they will promote stability at a time when the system is likely to be subject to considerable stress—some of which will originate from the international asset holdings of countries like Saudi Arabia.

Notes

1. No attempt is made to estimate government revenues from new exploration contracts, nor to forecast other oil company expenditures, such as wages to Saudi employees and purchases from local producers.
2. *Wall Street Journal,* October 19, 1973, p. 2.
3. *Petroleum Intelligence Weekly,* May 14, 1973, p. 7.
4. U. S., Congress, Senate Committee on Interior and Insular Affairs, *Oil and Gas Imports Issues, Hearings Pursuant to S. Res. 45,* 93d Cong., 1st sess., 1973, pt. 3, p. 921.
5. James E. Akins, "The Oil Crisis: This Time the Wolf is Here," *Foreign Affairs,* April 1973, p. 480.
6. "How the Arabs Plan to Spend Their Riches," *The Economist,* May 5, 1973, p. 39.
7. *Oil and Gas Journal,* December 25, 1972, pp. 82–83.
8. Senate, *Oil and Gas Imports Issues, Hearings,* 93d Cong., 1st sess., 1973, p. 1078.
9. *Oil Daily,* July 30, 1973, p. 3.
10. *U. S. News and World Report,* October 8, 1973, p. 95.
11. *Petroleum Press Service,* February 1973, p. 42.
12. Saudi Arabia Monetary Agency (SAMA), *Annual Report,* 1966, 1971.
13. International Monetary Fund, *IMF Survey,* July 9, 1973, p. 207.
14. In terms of dollars, the increase was 83 percent, reflecting the approximately 8 percent devaluation of the dollar in December 1971.
15. SAMA, 1971, p. 9.
16. Faisal Saffoq al-Bashir, *An Econometric Model of the Saudi Arabian Economy: 1960–1970* (unpublished Ph. D. dissertation, University of Arizona, 1973), chapter 6.
17. *Ibid.,* p. 114.
18. *Ibid.*
19. *Ibid.,* p. 115.
20. *The Economist,* May 5, 1973, p. 42.
21 *Ibid.*
22. *Newsweek,* July 16, 1973, p. 19.
23. *Wall Street Journal,* August 3, 1973, p. 1.
24. Akins, "The Oil Crisis," p. 481.
25. "Saudi Arabia: A Nation We'd Better Get to Know," *Forbes,* February 15, 1973, p. 30.
26. *The Economist,* May 5, 1973, p. 43.
27. "Statement by Assistant Secretary Joseph J. Sisco before the House Near East Subcommittee," *Press,* Department of State, No. 197, June 6, 1973, p. 5.
28. *The Economist,* May 5, 1973, p. 43.
29. SAMA, 1971, p. 62.
30. *The Economist,* May 5, 1973, p. 39.
31. SAMA, 1972, p. 39.
32. *Ibid.,* p. 36.
33. *The Economist,* May 5, 1973, p. 44.
34. Akins, "The Oil Crisis," p. 482.
35. *IMF Survey,* July 9, 1973, p. 207.
36. *Wall Street Journal,* July 27, 1973, p. 7.
37. National Petroleum Council, *U. S. Energy Outlook,* December 1972, p. 246.
38. Chase Manhattan Bank, *Capital Investments of the World Petroleum Industry,* 1971, p. 25.
39. *Federal Reserve Bulletin,* July 1973, A79.